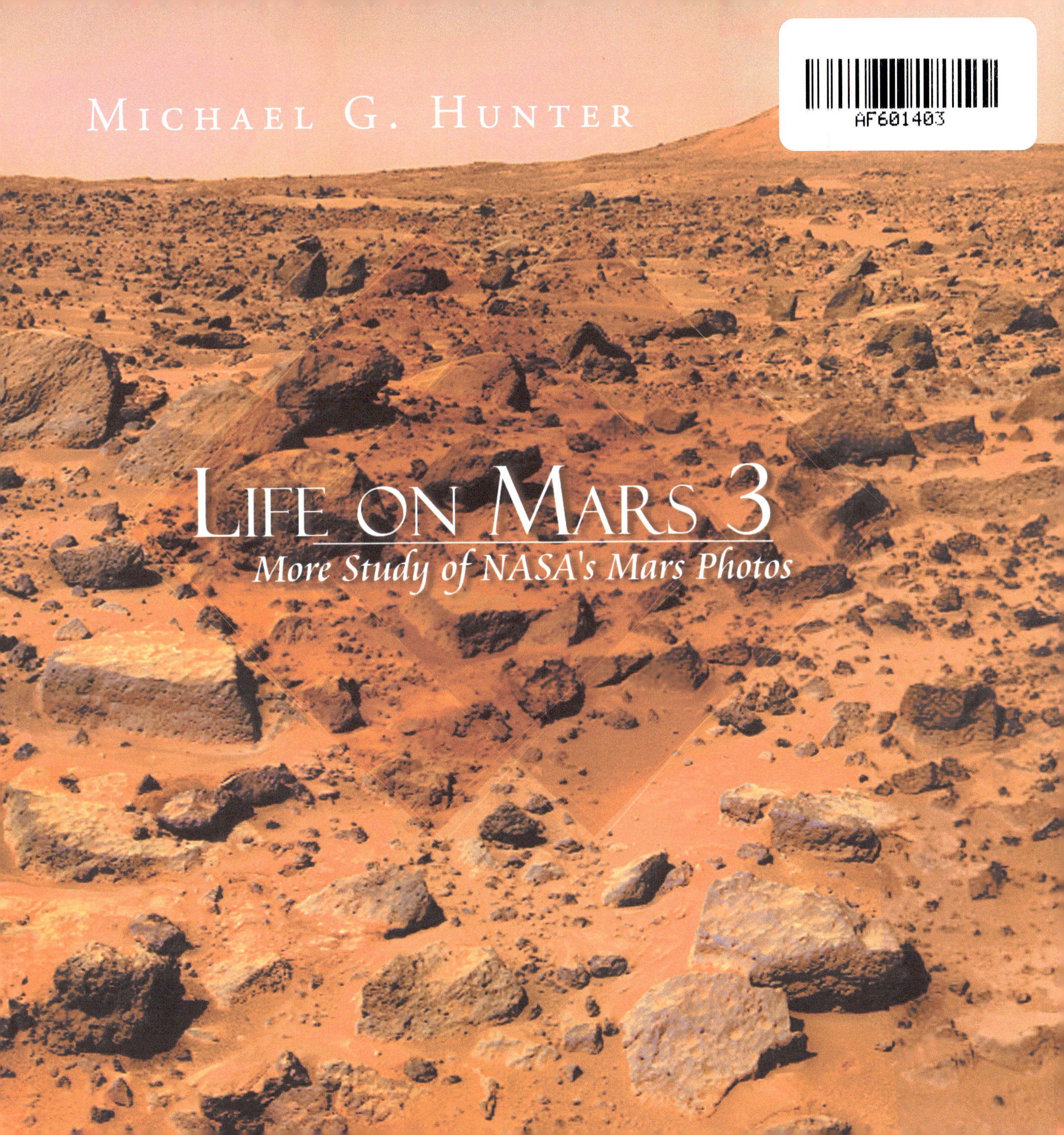
MICHAEL G. HUNTER
LIFE ON MARS 3
More Study of NASA's Mars Photos

ISBN: Softcover 978-1-4836-8438-3
EBook 978-1-4836-8439-0

Rev. date: 08/26/2013

To order additional copies of this book, contact:
Xlibris LLC
1-888-795-4274
www.Xlibris.com
Orders@Xlibris.com

LIFE ON MARS 3

MORE STUDY OF NASA'S MARS PHOTOS

Author
Michael G. Hunter

Editor
Theresa Brown

INTRODUCTION

This book is third in a continuation of a documentation of findings in a study of NASA's Mars rover photos. "Life on Mars" (published November, 2012) and "Life on Mars 2" (February, 2013) presented the photographic evidence of about forty species of animals as well as humans that have been photographed by NASA's Mars rovers and spacecraft. Those two previous books also contain several theories that help to understand how humans got on Mars, how and where the deluge happened, and how it might have happened more than once, even before the days of Noah's ark, in forgotten ages. There are essentially four theories brought to light in this series of books:

1) The theory that a Sun-partner (like Nemesis), was responsible for modifying Mars' orbit to a more elliptical shape on a regular basis, every 37.5 million years, which caused it to interact, brush, or collide, with Earth;

2) The theory of Mars-Earth exchanges, in which life forms from Earth and Mars interchanged between the planets Mars and Earth many times, including the most recent Great Deluge of the Bible, during the days of Noah and his ark;

3) The theory that there are scars on Earth and on Mars that resulted from those many collisions; and,

4) Explanatory theories of life on Mars, e.g. how Martians survive, how water is sustained, how plants and animals are able to survive and thrive in the seemingly harsh environment.

Early man didn't really know the order of the Universe as we do today. The Sun appears to pass across the sky in the same manner as the Moon, planets, and stars, giving it the illusion of revolving around the Earth. Nevertheless, as early as the 3rd century, BC, the Greek astronomer, Aristarchus of Samos, wrote about a universe with the Earth travelling around the Sun, and stars as distant objects too far away to see movement. Independent thinking astronomers such as the Indian astronomer, Aryabatha in 499 AD, wrote about that concept as well. A thousand years later, when Nicholas Copernicus wrote, in 1533, that the Earth revolved about the Sun, the fact was utterly rejected by the great authority, the Holy Catholic Church. Only after Isaac Newton's brilliant scientific writing in the late 1600's, and their broad acceptance, did the authority finally come to agree with the reality of heliocentricism. Until then, the concept was considered heretical and blasphemous, because it simply contradicted the Bible. The concept of a body travelling through space with no means of support was difficult to comprehend in the days before modern science; and that the world, Earth, was no more than a tiny speck of dust amongst a boundless cosmos, was unfathomable and out of sync with the religious teachings.

Nowadays, in a similar circumstance, belief of life on the waterless, cold, oxygen deficient planet Mars is currently disregarded by scientists, and will probably continue to be, until it is proven with biochemistry or some other modern scientific discipline. But some people commonly say, in apparent conflict with modern science, that "seeing is believing", and "photographs do not lie". Thus, this book is an attempt to show that the scientifically unbelievable can in fact be true, that science has its limitations, and that mankind in his modern Earthly world is and has been blind, deaf, ignorant, miscalculating, and misjudging about Mars, not just in the distant past, but in recent times and even today. The concept of life on Mars has become heretical and blasphemous, contradictory to scientific theories, and beyond belief, a concept which this book shows to be flawed.

Curiosity rover, which landed at the outset of the first book, (and partly inspired it), during this writing, is still working its way through Gale Crater, on its way to Mount Sharp, drilling and analyzing rocks at Yellowknife Bay, where it seems to find, appropriately, liquid water in pools, with animals seemingly taking advantage of the last vestiges of water as it is about to evaporate forever in the cloudless, rainless, nearly waterless world with meager atmosphere. The scientific investigatory device suffered memory problems in late February, (computer malfunction), but fortunately had a backup computer for that and proceeded. Then again, in April and May, Mars passed behind the Sun, again causing delay in its progress.

Scientists are developing an understanding that Gale Crater (and Mars elsewhere) must have been at one time full of water, which, (we can see from the photographs), is now evaporating and changing from a lake to a mud hole.Yet, in spite of this world of vanishing liquid water, NASA seems lost in their world of rocks and geology, unable to see the water, dinosaurs and strange animals and humans, that live all around the rovers. Either “mum’s the word” or the thousands of NASA employees are simply inattentive to the photos.

The old adage, “Seek and ye shall find,” is at the heart of the subject of life on Mars. It’s a matter of attitude. If you look at a photo with the objective of seeing life forms, you will see it, whereas, if you reject the idea outright you will not. If you do not seek, you will never find. This is the reason why life has not been seen on Mars for the past fifty years in spite of photographs that show otherwise.

The ancient written accounts are murky, scanty, vague, confusing, and difficult to comprehend. People today are ignorant of the whole flood event, wondering why civilizations like Atlantis disappeared in a day, wondering why there are sand-covered cities older than the flood, (for example, Göbekli Tepe, in Turkey), wondering why structures now can be found beneath the oceans and at the bottom of lakes. People search for answers all over the world, wondering, searching, trying to understand unwritten human history. This book joins the effort and hopefully adds some answers.

One would think that the primary concern of humans living on Mars would be the low temperatures, the suffocating oxygen level, the lack of water, or perhaps the relatively miniscule atmospheric pressure. However, in the first two books, I have revealed a society of humans on Mars that seems lost in time, five thousand years ago, yet is likely most complicated by the issue of SIZE. In "Life on Mars", giants sixty feet tall were seen to live alongside of normal people, yet they keep out of each others' ways, or so it seems. In "Life on Mars 2", Super giants, hundreds-of-miles tall, were revealed, who also seem to keep well out of the way of the smaller inhabitants, staying in their boundaries of craters; for one misstep with a ten-miles-long foot could effortlessly crush and destroy a city, squashing its inhabitants like Earthlings trample ants. Now, in this third book, yet another complication is revealed...human miniaturization. I present the little people, tiny little humans, the size of your little finger. This discovery reveals and further complicates the major difficulty of Martians, i.e., protection from accidental missteps. Mars is a bizarre world compared to Earth. People on Mars not only grow to enormous size, they grow smaller as well. Yet they live together on a planet half the size of Earth. My unbelievable story that started with a simple sand dollar in "Life on Mars" now seems to be extending endlessly into a surrealistic world, similar to the fictional account of "Gulliver's Travels", with tiny little humans sharing a planet with giants and animals, strange beyond imagination. These tiny people did not shrink suddenly like in the movie, "The Incredible Shrinking Man". It happened, more likely, over many generations; and likewise, the giants grew similarly, over thousands of years. And, during the last collision, in the time of Noah, they fell to Earth, giving us myths of tiny people, "pixies", and gigantic "gods", who fell from the sky. These abnormal-sized humans will undoubtedly lead to difficulties with Earthlings in the future, as we Earthlings occupy Mars, in the coming decade. We will need to tread lightly.

One thing that inspired me to write this book was the fact that nobody seems to believe that there is really Life on Mars. Schools teach us that Mars is a dry and uninhabitable, lifeless planet, and internet websites such as NASA's and Wikipedia's as well as public television documentaries, all give support to that fallacious concept. I want to let the world know, and I want to set the record straight, that there IS life on Mars...and there ARE humans on Mars. These books I am writing present the evidence, for everyone to judge for themselves. They show dozens of animal species on Mars, as well as four general sizes of humans, including the tiniest, the normal-sized, the giants, and the super giants. Awaken, dear fellow Earthlings! We have interplanetary companions! that we have been subconsciously yearning for. Here are some photographs of them.

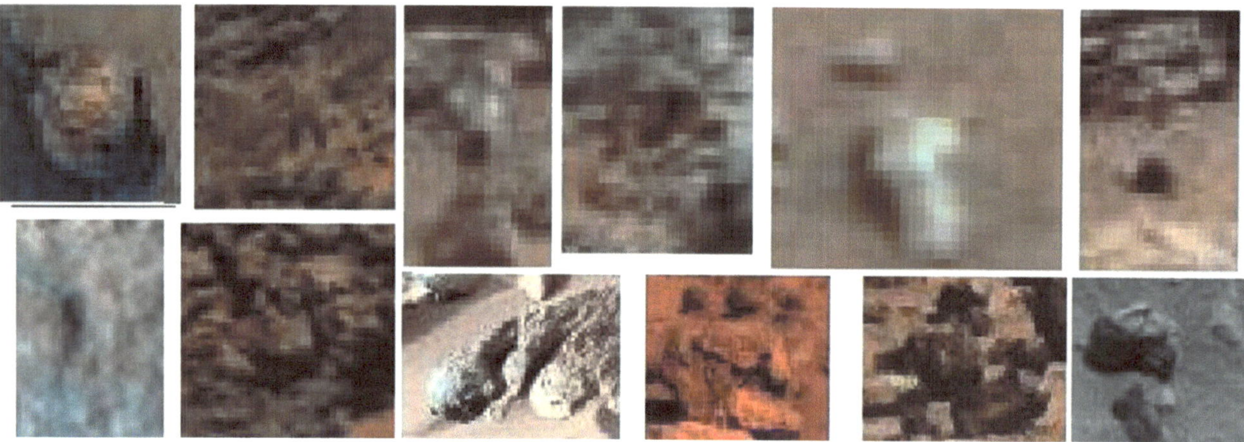

PART 1: WATER ON MARS

THE TWO GREAT FLOODS

In the beginning, God created the heavens and the Earth. "And, on the first day, he created light which shone upon the waters". Therefore, based on this wording from the Bible, waters existed before anything else. One interpretation of that mysterious and confusing writing of long ago is that there was a flood before the great flood. Following the first flood, according to the Bible, the surviving humans were Adam and Eve, and their offspring became human civilization on Earth. Cities and entire societies that developed prior to what we now call the Great Deluge, were destroyed by the second flood in Noah's time, and buried or sunken, then forgotten but for vague ancient mentions. Archaeologists lately are now finally, after several millennia, finding lost civilizations all over the world: Ancient Egypt, Gobekli Tepe in Turkey, Puma Punka in South America, Yonaguni off the coast of Japan, Bimini Road in the Bahamas, and so on, civilizations that existed before the Great Flood. On Earth, today, we seem to have all but forgotten about the Great Flood(s). Almost everyone who was able to witness the flood firsthand probably drowned. Others who were not able to witness it directly probably did not know what happened or how, because the sky was obscured by clouds, storms covered up the stars, rain came down in bucketsful constantly, while winds and waves tore apart houses and blew people hither and thither. But someone DID survive, here and there all over the Earth, people on boats, and on high mountain peaks. What people could not see then can be seen now in mankind's curious mind: water, a great tide, was sucked by Mars' gravity to one side of the Earth, with a three-mile high wave of water. People who survived passed knowledge of the flood down by word of mouth, through many generations, leaving behind what we know today as vague myths and legends, all over the world. Noah did not survive the flood alone with his three

sons and their wives. In the Andes of South America, according to Inca myths, someone survived by hiding in a cave near the mountaintops, only to discover later that he was one of the few lucky ones to survive, in his vicinity. All over the world, in other continents, in Asia, in Africa, in the Americas, and in Europe, people survived the flood. The fact that today we have people of many races, shows that people, not just Noah and his family, DID survive. There are records of humans that go back tens of thousands of years. We can see that now better than ever before, through various scientific advances such as radiocarbon dating and DNA research. There were societies before the Great Deluge. And there were probably prior deluges.

THE GREAT FLOOD ON EARTH

When the mythical Great Flood occurred on Earth, water level on Earth rose, in the country now known as Turkey, to the top of Mount Ararat, a mountain that rises over 13,000 feet above sea level. Noah's Ark has been found there recently, confirming that event. Millions of people must have drowned during the calamity. As Mars collided with Earth, a four-thousand-mile-diameter globe with countless numbers of people living on its surface struck another globe twice its diameter. Both were travelling in their orbits around the Sun at about 60,000 miles per hour, while rotating at 1000 miles per hour at the equators, scraping together like grindstones. They collided in one of Earth's most populous regions, the Mediterranean Basin, causing an impact scratch 2500 miles long on both Earth's Mediterranean Sea and Mars' Valles Marineris. It is no wonder that the ancients considered it an act of God, punishment for sinful acts. After all, almost everyone in the area, on both planets, suddenly and unexpectedly died from a vast natural phenomenon. And historically, God had wiped out cities before, seemingly as punishment, with meteor strikes, and so people knew that God could be merciless if he wanted to be.

According to the Bible, the flood was an event manipulated by God to purposely eradicate mankind, leaving only Noah and his family, along with the numerous animals that were taken aboard his ark. Noah's ark was found, with 99.9% certainty (claim the discoverers), at elevation 13,000 feet on Mount Ararat, in 2008. Radiocarbon dating put it at 4800 years old, i.e. about 2700 BC. However, Noah was not the only person with a boat at that time. I suspect that anyone with a boat, when the floodwaters rose, would have gotten on their boat when the floodwaters approached. In Ancient Egypt, for example, boats floated the Nile all the time. And a flood in Turkey, thousands of miles north, does not necessarily mean death for all on the Nile, nor in southern Africa. Some animals can swim, and likely did, when the waters came. Some animals live under water, like fish and mollusks. And we know that people and animals fight for life in catastrophes like the great flood, powerful earthquakes, hurricanes, meteor and asteroid collisions, and other calamities, by clinging to floating objects, hiding in high mountain caves, or whatever means is available. Animals, people, and the artifacts of society were washed onto the four-thousand mile diameter red planet, as a great wash of water with its soup of creatures, from Earth was splashed onto it, while it was spinning and passing by at thousands of miles per hour. Entire civilizations were ground to oblivion, (like Atlantis?).

THE GREAT FLOOD ON MARS

The deluge probably caused worse misery on Mars than on Earth. The rover photos show that the Martians still have not recovered, five thousand years afterwards. They still haven't cleaned up the debris from the deluge, as Ancient Egyptian artifacts and ancient boats, unable to rot in the thin air, still lie on the ground like in dried-up lakes on Earth. The water has mostly evaporated after five thousand years, but there are still signs of the great wash, with age old streams and riverbeds abundant, as we can now see with the rover cameras. Vast landscapes are littered with debris from the great wash of water from Earth to Mars. Boats with and without people aboard, animals floating on logs, even stone sarcophagi with bodies inside, anything that floated, washed ashore onto Mars

as it collided with Earth. Undoubtedly, if there were people on Mars, and it appears to be so, people drowned or died on Mars from the impact with Earth. Some people from Earth seem to have been transferred, washed to the passing planet, and they survived. For, not only water washed to Mars, but atmosphere as well, which, like the water, is slowly fading. The survivors are now Martians: however, Earth was their home planet. And over the eons, they have transformed and adapted to the strange world with a dreadful destiny. Unlike those on Earth, Martians may have been able to see the approaching planet and catastrophe coming to them, because they did not have clouds obscuring their vision, and so they may have ancient writings or legends about the event.

BOATS ON MARS **FROM PATHFINDER 1997**

The with its soup of living creatures, above images appear to be the hulls of boats, stranded on the desert, possibly remnants from the great deluge, five thousand years ago, (or at least, a sign that water was once much more plentiful), from the 1997 Pathfinder image used on the cover of "Life on Mars".

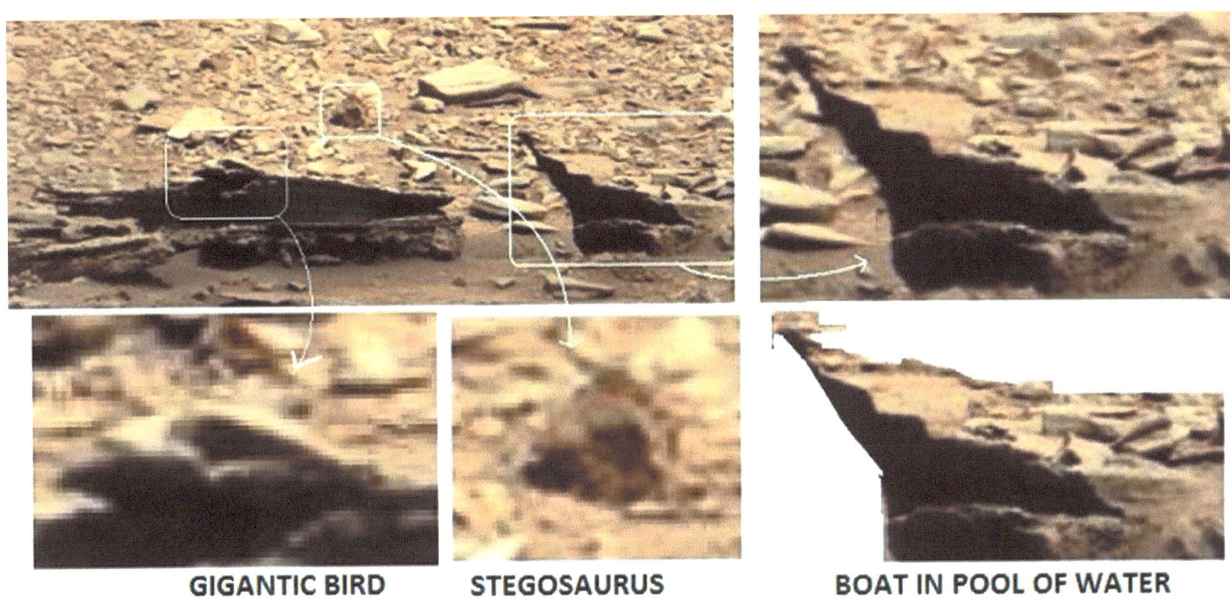

And there is this image from the Curiosity rover, showing a boat in a pool of water. It could be a remnant from the flood in Noah's time...or it could be more recent. The water appears to be liquid, not dried up nor ice. The boat appears to be floating in the water, not sunken. The huge bird next to it appears to be larger than the boat. The scale on this photo indicates that this boat is only about one foot long. Therefore, this boat must have been used by tiny people in the days of great water on the planet. Note, also, from the same photograph, that the stegosaurus, a huge animal on Earth in times long past, seems tiny in relation to the gigantic bird.

ANOTHER DAM!

DAM WITH WATER BEHIND IT

The photo above shows another dam, (similar to the one noted in "Life on Mars"), but this one is taken from above, and shows liquid water behind it. Notice in the lower photo how the dam is constructed with gravel and sand, in a perfectly straight line, similar to the other dam in Gale Crater. The water level is slightly below full capacity, indicating that it may be nearly full and probably built recently. Considering that there are no clouds and there is no rain, and the water would evaporate quickly in the low atmospheric pressure, the water must be draining to the dam on a continual basis. Having seen two dams on Mars suggests that water is indeed a matter of concern to Martians, and that they do exercise effort to control it for their survival. Dams on Earth are built for similar purpose...to collect water from a large area so that it can be utilized, but also to alleviate water shortages.

The construction of dams, both the one in "Life on Mars", and this one, is similar. They seem to be simple earthwork dams, without any concrete like modern dams on Earth. This says something about the technical capability of the Martians. They are not apparently using concrete to a significant extent. They do make metal pipes, bronze-looking. It could be that they are still in the Bronze Age, not yet knowing about steel. And, they do make earthen structures, such as dams and roads. It's an odd mixture of ancient and modern. People ride elephants and camels and horses, while others have cars and even jet planes, (as seen in "Life on Mars 2").

WATER IN YELLOWKNIFE BAY

Since late December, 2012, Curiosity Rover has been in the area they call Yellowknife Bay, photographing geological features and performing scientific studies. Although NASA added captions to the photographs relating to the geological features, they did not mention the pools of water in any of the photographs. They stare into the eyes of dinosaurs, neck deep in mud, and call them rocks. They maneuver the rover around water pools, yet see only stones. Perhaps they see the water, but just do not wish to get into that subject now. They are so determined and focused on finding microscopic life forms, they do not wish to be bothered by claims of animals and people. They cannot see the forest through the trees. The water must freeze at night, then thaw at daytime, at least on the surface. Therefore, for those who cannot see or believe in Mars' water, read on.

POOL OF LIQUID WATER UNDER A ROCK **CURIOSITY DECEMBER 2012**

This December 24, 2012 Curiosity rover picture, above, shows a pool of liquid water that seems to be percolating or pooling from under a rock.

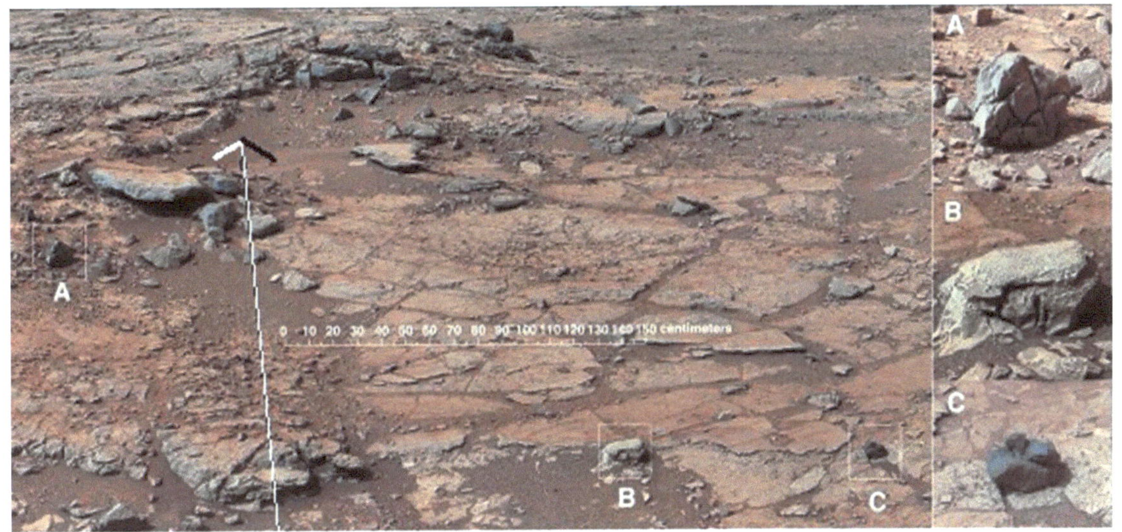

POOL OF LIQUID WATER **CURIOSITY DECEMBER 2012**

Another photo, above, was taken December 25, 2012, showing a small pool of liquid water.

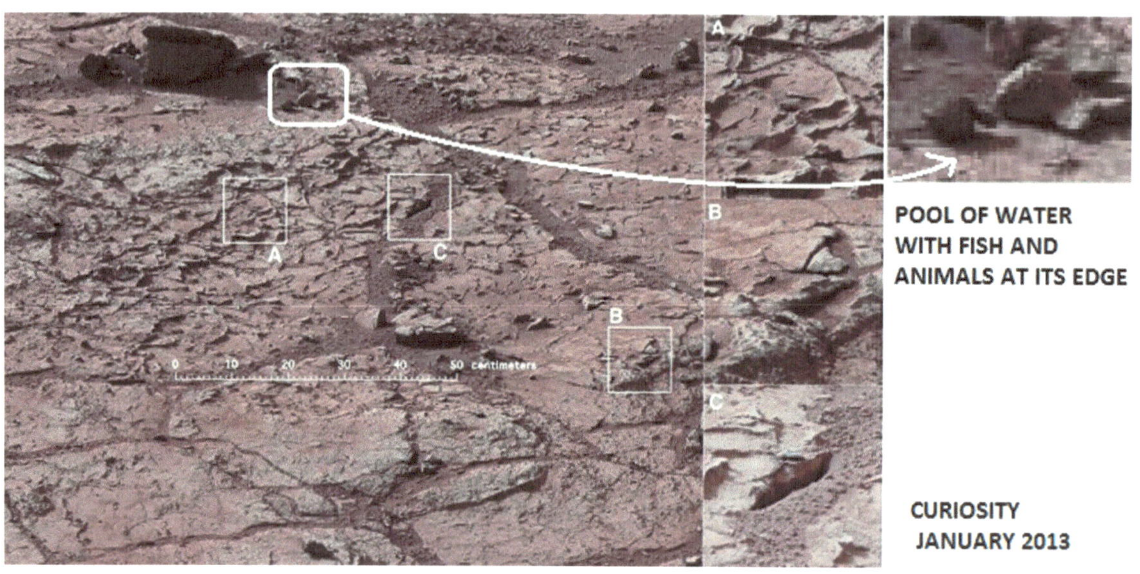

January 10, 2013, above, a drying pool of water with fish and other animals making what they can of the water as it looks like it is soon to evaporate totally.

At a site called "John Klein", on February 3, 2013, notice that next to the rover is a pool of liquid water. The Curiosity rover drilled into a rock at this site.

POOL OF LIQUID WATER CURIOSITY JAN 2013

February 8, 2013. This photo, above, shows how the rover drilled the surface. Notice the pool of liquid water, and Mount Sharp in the distance, the rover's destination.

Why so many pools of liquid water at Yellowknife Bay? Well, after all, it has been named a "Bay". The rover is approaching the huge lake in Gale Crater's center, and similar to a lake on Earth, it has tributaries that drain into it from the surrounding area. There must be subterranean ground water

draining into the lake. This might be water from long ago, or it might be water that condenses daily on the surrounding surfaces.

In summary, there seems to be liquid water on Mars, contrary to current belief. The "dry" planet is not so dry. The water probably melts during the daytime and during summertime, but is frozen during the cold nights and during wintertime. Many photographs from rovers, including Opportunity and Viking Lander rovers, show that liquid water does exist on Mars in diverse locations. Although it is relatively difficult to see in rover photos, it apparently is sufficient to sustain life of animals and humans in many places throughout the planet. Also, we cannot see beneath the sand, so it is possible that there are vast quantities of water just out of sight.

PART 2: PLANTS ON MARS

There is no rain on Mars, so that pretty much keeps what we Earthlings consider plants, i.e. green chlorophyll life forms, from growing, at least on most of the seemingly dry world of Mars. There is sunshine, although it is significantly less strong than on Earth, due to distance from the Sun. The Sun's distance to Mars is roughly 1.6 times that from Earth, (150 million miles / 93 million miles =1.6), and since the intensity of radiation decreases in a ratio disproportionate to the distance-squared, sunlight is 1/1.6 x 1.6 or about 40% as intense as on Earth, without considering atmospheric influences. But plants cannot live on sunshine alone. Photosynthesis is the process in which plants use light energy to combine water and carbon dioxide to form carbohydrates while giving off oxygen. Carbon dioxide on Mars is plentiful for the process, but everyone knows that without water, plants don't grow. From the previous part of this book, we know there IS water, along with sunshine and carbon dioxide on Mars. Also, we know that temperature affects plant growth. Frost kills all but the heartiest plants in wintertime, so a night temperature on Mars of -150 degrees F is not conducive to plants. That alone probably explains why landscapes of Mars appear barren and plantless. Yet animals, plant-eating types such as herbivorous dinosaurs, ducks and geese, bighorn sheep, giant panda, camels, hippo's and elephants seem to be alive and well on Mars, suggesting that plants DO exist on Mars, somewhere, and somehow, as yet, unknown. All of those plant eaters need to eat SOMETHING, but what is it they are eating and where is it and why can't we see it? Is there some red algae or some kind of food source under the sand? This question will remain unknown until we get better visibility or information. Not only do animals and people seem to go about on Mars in great numbers, in spite of the subzero temperatures, and negligible oxygen levels, but they seem to live without a visible food source. It's just one great unsolved mystery on top of another. Animals and humans on Mars live without plants, and in freezing cold and hardly any oxygen or atmosphere. Yet, there they ARE. It seems impossible.

POSSIBLE GREENHOUSE

The pictures below show what looks like a greenhouse, photographed from an orbiting satellite with telescopic camera. People can be seen inside and outside the greenhouse. A lamp is held by a woman inside, as I have diagrammed. Is it an electrical device she holds or a candle? Obviously, the oxygen level does not support a candle so it must be something else. There appear to be flowers like petunias in the greenhouse. A man can be seen entering the greenhouse, and another man and child can be seen nearby. But even though there might be some greenhouses here and there on Mars, it just does not explain how all the animals and people survive, seemingly everywhere.

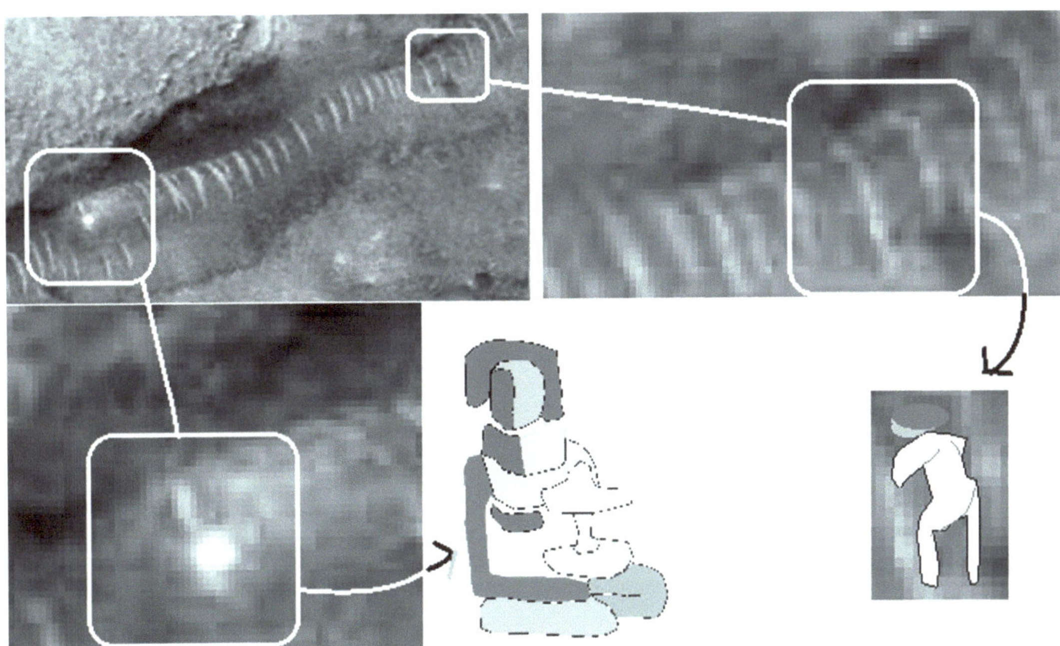

The photo below is a view of the overall structure, a possible greenhouse built by Martians, made with some kind of bent flexible wooden boards and a transparent or translucent covering, like polyethylene sheets. The sawn wooden boards suggest that they have trees somewhere in the vicinity, and also that they are able to saw them with some kind of crude device.

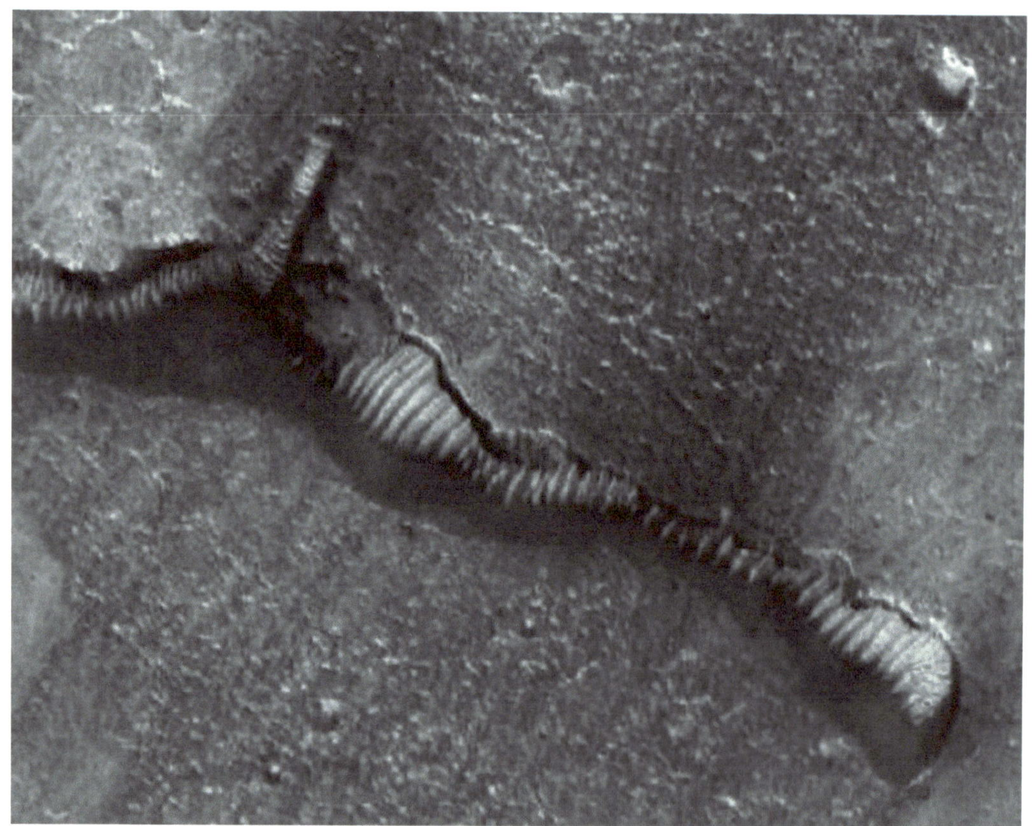

TREES

Other photos show images that look something like trees, although not exactly the same as those on Earth, like in the following satellite photo.

MUSHROOMS

Is there a type of plant that survives without light and without oxygen? Did you ever hear the old saying, "I feel like a mushroom; they just keep me in the dark and feed me b.s." That leads to the mushroom theory. Maybe there is a layer of fungus under the surface of the soil, subterranean. Animals are subterranean, why not plants or fungi? Plants on Mars either don't exist or live out of sight.

Fungi include yeasts, molds, and mushrooms. They constitute what is called a "Kingdom", separate from the plant, animal, and bacteria kingdoms. Lichens, shown in "Life on Mars 2", to exist on Mars at the Pathfinder site, are yet another type of life outside of those kingdoms. Genetically, fungi are more like animals than plants, having diverged from both about a billion years ago. Mushrooms are edible, as most everyone on Earth is aware. There are possibly 5 million species of fungi on Earth. Some are poisonous. They have roots and spread by their roots, as well as by spores.

So, perhaps, there are fungi under the soil, which we cannot see, but are plentiful, available to vegetarian animals.

GIANT PETUNIAS

As I mentioned above, there appear to be petunias in the greenhouse above. I have also seen what look like gigantic petunias, elsewhere in photos, very difficult to discern due to the poor photography. In the first Curiosity photo there appeared to be pansies, and in the Spirit Opportunity photo there was a petunia-looking plant, mentioned in the first book, "Life on Mars". The petunias I have seen appear to be gigantic, similar to the giant people. The rover photos do not show them clearly, and their edibility seems unconvincing. Furthermore, there are no pictures of animals feeding on them. Therefore, it is unlikely that they are a food source.

INVISIBLE PLANTS

The idea of a world without plants seems unlikely. Many, if not most, of the animals that have been detected in this study are plant-eating on Earth: stegosaurus, the hippopotamus, the ducks, the geese, the giant panda, elephants, camels, cows, horses, and the bighorn sheep. It does not seem possible that all of those animals exist on Mars and yet became carnivorous, with large animals eating small ones. So the inescapable conclusion is that there are plants, of some kind, on Mars; however, they are invisible to our rover cameras. We know that some photographs show that flowers and lichens and mushrooms do exist in the harsh, cold, dry environment. But where are they and what are they? For some unknown reason, we cannot see them.

The logical assumption is that they are there, but invisible to us. They are simply hidden from view. On Earth, some things grow beneath the surface of the soil, such as tuberous plants like potatoes, carrots, turnips, and radishes. But we can easily see their leaves above the ground, and without the leaves, the tubers would not grow. Also fungi like mushrooms grow in dark places and might grow under the sandy soil on Mars. Another type of possible food source is algae. And still another possibility is seaweed, assuming there are pools of water beneath the surface, which cannot be seen by the rovers. It seems the rovers have not explored the areas beneath the sand, probably because they need to avoid sand pits, due to the possibility that they could become stuck. Without being able to see them, we can only make logical assumptions that they are there. If not, how can we explain the herbivores?

GREENHOUSE EFFECT

Is there a greenhouse effect on Mars? On Earth, we are concerned that the buildup of carbon dioxide will lead to global warming which can cause species extinction. On Mars, there is a vast amount of carbon dioxide in the atmosphere, (96 percent). Does it help to moderate the temperature beneath the meager atmosphere and help to block cosmic radiation?

KELP

What we see in pictures looks like desert. However, many photos show shallow water pools. Also, there are photos that show people in the sand, up to their waist, or bathing or washing clothes in the sand. There is likely water beneath the surface! If so, it seems possible that there is plant life below the surface, as well, hidden from view, similar to the way kelp always stays below the surface of water, on Earth. We can see kelp beds on Earth because water is transparent, but, perhaps there are invisible but vast quantities of plant materials, such as kelp, beneath the surface of the sand, just out of sight to our rover cameras.

PART 3: ANIMALS ON MARS

"Life on Mars" and "Life on Mars 2" presented, in total, about forty species of animals, visible clearly enough to be identifiable, on NASA's rover photos, including:

EARTHLIKE ANIMALS

sand dollar	rat snakes	giant panda	stegosaurus
ducks	weasel	mourning dove	velociraptor
snow geese	bighorn sheep	hippopotamus	brontosaurus
elephant	cow	dogs	horse
cow	moose	Pterodactyl	Red fox

NON-EARTHLIKE ANIMALS

Super geese	Super-duper croc	Rock Mouths	Turtocerouses
Super raccoons	Super dog	Broken Rock Bunnies	Mantis predecessor
serpopard	jake	lizards	Giant boa

MORE ANIMALS ON MARS

ANIMALS OR ROCKS?

This looks like a pile of rocks. But notice how its head is symmetrical, with two large bulging eyes, and a beak? Also a young one is at its side. It appears to have hidden itself in plain sight by putting together rocks and then camouflaging itself to the same color.

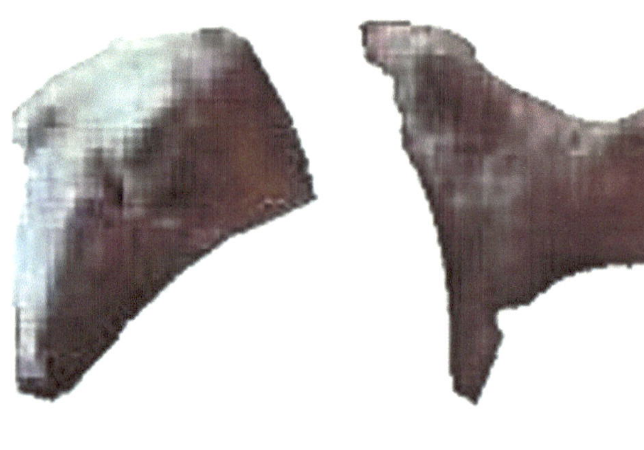

ROCKPILE ANIMAL

This is another similar rocklike animal. It seems to have feet that elevate it above the ground, as well as a rocklike shell, a beak and bulging eyes. NASA considers this an "interesting conglomerate".

These seem like rocks. But why are they shaped like animals?

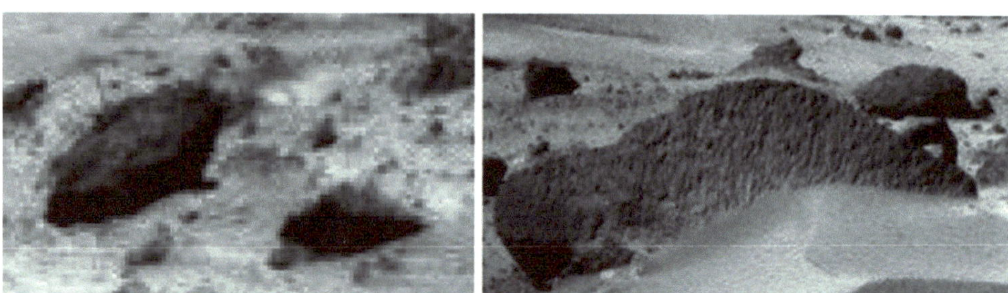

ROCKLIKE ANIMALS

EARTHLIKE ANIMALS

Below left is a Curiosity photo of an elephant with a rider; notice cylindrical tank beyond the rider. The elephant seems to be walking in a pool of sand.

ELEPHANT WITH RIDER

STEGOSAURUS

Above right is another photo of a stegosaurus from the Curiosity rover. This is in addition to the few photos of stegosaurus in "Life on Mars".

The photo looks like a bird spreading its wings. Some birds on Earth do this to dry out in the sun, e.g. cormorants.

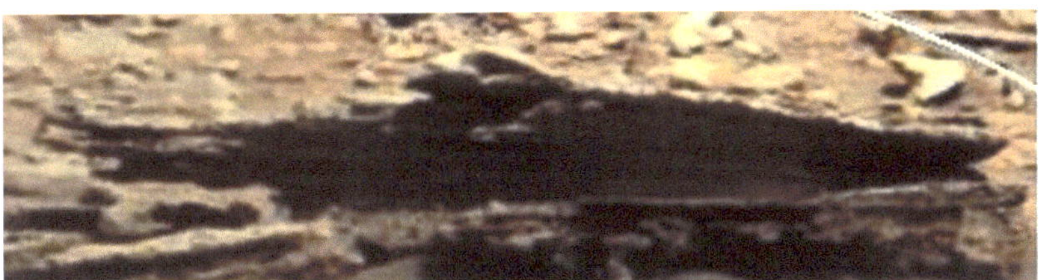

BIRD SPREADING ITS WINGS TO DRY

The photo below shows what looks like a brontosaurus, head pinned to the ground, with man standing next to it. A second brontosaurus is beyond. Notice also the possible car in the background, very much like a car on Earth from the 1950's.

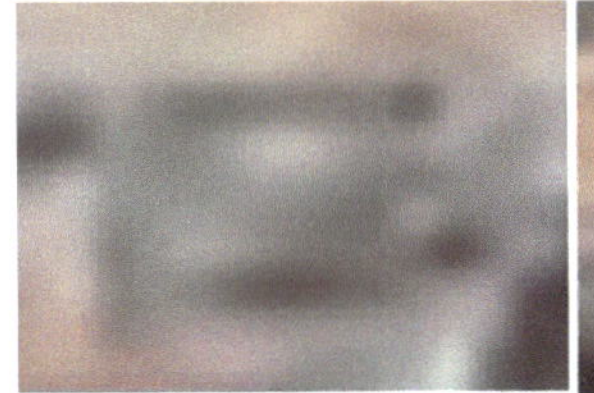

CAR

BRONTOSAURS WITH HEAD STRAPPED TO GROUND WITH MAN POSING BESIDE IT

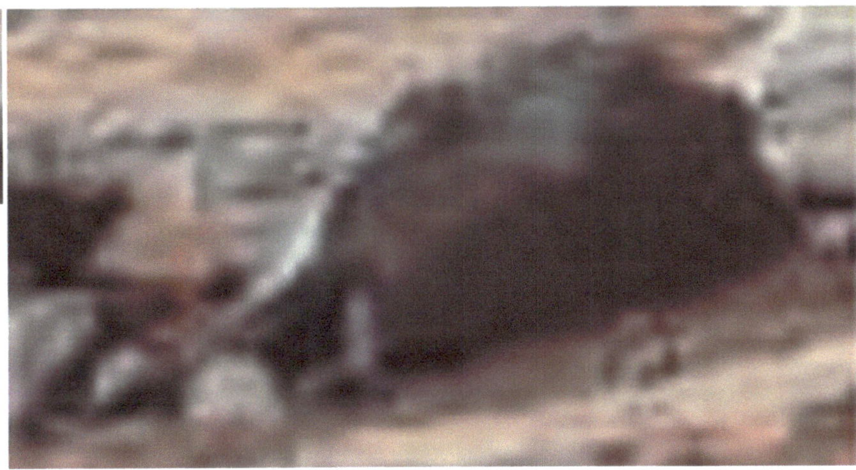

The following photo appears to be a man mounted on the back of a camel, about to stand up.

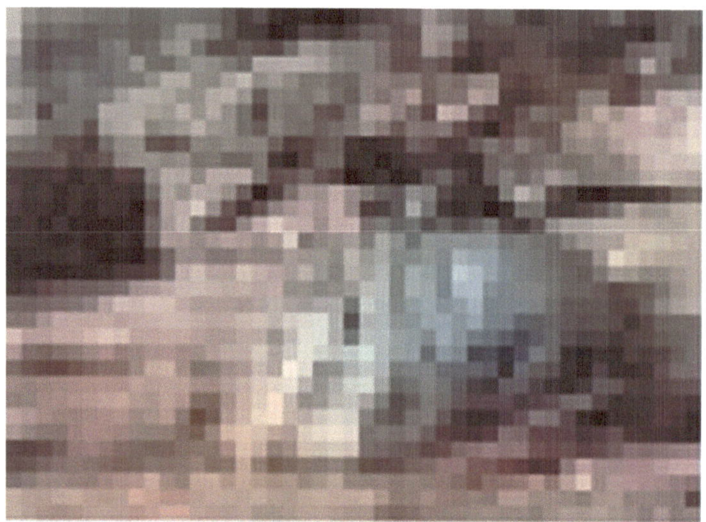

MAN MOUNTED ON CAMEL

PART 4: MARTIANS

THE TINY PEOPLE

In addition to normal, giant, and super-giant humans, NASA's photos show that there are tiny people on Mars. The Spirit rover photo, below, was taken on April 13, 2006, with a tiny little man standing on a rock, holding his hand up in the universal "greetings" gesture. This shows that these little ones exist and would like to contact us. NASA claims that this boulder is about 16 inches tall. On that basis, the little man measures 1 inch tall. Upon careful study, he seems to be wearing a cowboy hat and something like a baby carrier on his chest. The color of his clothing seems to be blending in with the rock, which is lava. Notice that he is not alone. Another such tiny person is at the base of the rock, standing on top of a smaller rock, with what appears to be a megaphone propped on the rock, next to him. He also has one arm raised, as if waving to the camera.

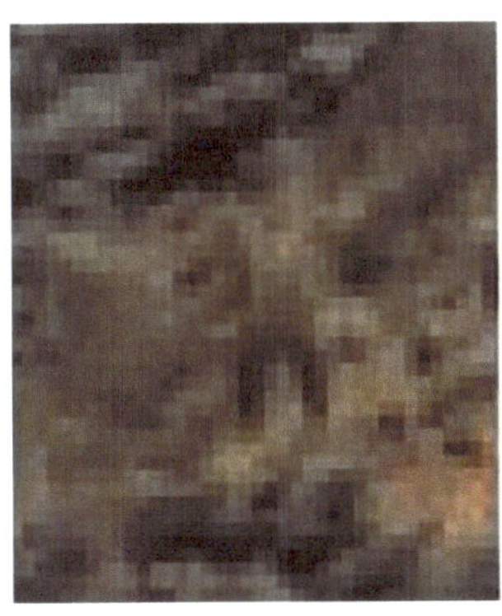

TINY MARTIAN WAVING TO OPPORTUNITY ROVER CAMERA

The above finding , showing tiny people, waving to the camera, is heartwarming and heartbreaking at the same time. It seems these people are truly trying to offer a gesture of peace; meanwhile, we are unable to see them clearly or respond in kind. We are essentially blind to them. It's probably very frustrating, for them, to have put out such a gesture, only to have it ignored.

The photo below, from the Curiosity rover contains what looks like a nice typical family. There is no way to see or prove or verify what the actual sizes of the people are, but there is a scale on the drawing. The little girl on a rock, by that scale, is about three centimeters tall...a little over an inch, (same as the men above). The elephant, lying under a slab of rock, would be a foot tall if standing up. The duck would be four inches tall, four times the size of the little girl. So many things seem out of proportion.

The photo shows a family including: a four or five year old lad in shorts standing on a rock with a ball in his hand, a man, probably the father, washing his face in a sand pit, a woman doing laundry in the sandy water, a teenaged girl in the sand up to her waist, a little girl, perhaps six to eight years old, wearing a black skirt with a pink blouse, dancing on rock, and the duck is in the sand, bigger than all

of them. There are other strange animals in the photo, such as what looks like an elephant under a rock slab, shooting a wad of some kind through its trunk, at a kind of giant dragonfly. Reptiles roam here and there. In the same photo that this was extracted, (but not in this enlargement), there is what looks like the hull of a boat sitting in a dried up pool of water. The boat seems to scale similarly to the humans...only six inches long. But next to the boat is a huge bird, spreading its wings, even bigger than the boat! See photo in Part 1, Water on Mars.

Below is another photo enlarged from a Curiosity rover image, from August, 2012, just after it landed, showing the ground beneath the rover. It's quite difficult to make out, but it does seem to show a group of several humans, sitting on rocks, facing the rover. There is also what I believe to be petunia flowers on the ground around them, as well as a vase of flowers sitting on a rock or table, on the far left side of the photo. As usual, the photography quality makes it almost impossible to tell precisely what the people are doing or what their facial expressions are.

TINY MARTIANS CELEBRATING CURIOSITY LANDING 8/6/2012 **MGH 8/15/13**

For yet another example of these tiny people, the photo below was taken by the Curiosity Rover on December 24, 2012. It shows a rural village, with all sorts of things going on. The scale demonstrates that they are just five centimeters (about two inches) tall, similar to the size of the previous examples. There are boats, waterways, farm animals, a brontosaurus, a camel, and quite a few people. One man seems to be juggling pots or performing some kind of balancing act for the camera.

TINY MARTIANS IN A VILLAGE

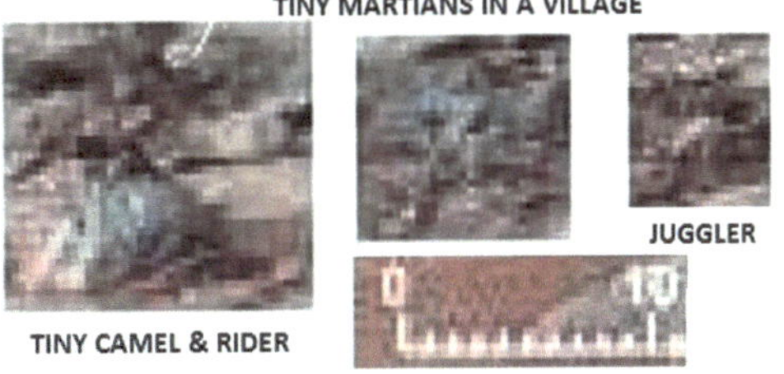

JUGGLER

TINY CAMEL & RIDER

SIZE DISTORTION

So, how did such people come to be so small, and similarly, how did the super giants get to be so large? There seems to be something about Mars that causes a distortion in size of animals as well as humans. It probably occurs over long time periods, over many generations. For example, in one area, the geomagnetic field is positive, or north, and in other areas, it is negative, or south. And the result apparently causes an increase or decrease in size of offspring, respectively, so that perhaps each generation shrinks or grows by ten per cent, or something like that. After multiple generations the people (and animals) become equally small (or big), in a local area or region.

STONE CARVINGS?

The photo below shows a rock with liquid water at its base, as noted in Part 1: Water on Mars. The rock appears to have carvings on it. This could be artwork that was made on Mars, or possibly, something that washed up to Mars during the Great Deluge. The work seems to depict human forms. One looks like a mermaid. Another looks like a face. Still others appear to have eroded beyond recognition. And of course, they could simply be natural formations. We need to explore them further in the future.

HOW DO PEOPLE SURVIVE ON MARS?

The atmosphere extends above the surface of Mars about ten miles, compared to about one hundred miles, on Earth. The atmospheric pressure on Earth at its highest point, the top of Mount Everest, about 10,000 meters above sea level, is three times that of the lowest elevation on Mars. The oxygen level on Mars is just 5 percent of that on Earth. However, mountain climbers have gone to the top

of Mount Everest, usually with oxygen tanks, yet the local Nepalese guides, who grew up in the high altitude, apparently do not need the oxygen. Perhaps, with proper acclimation, people from Earth could develop ability to withstand the low pressure and low oxygen levels on Mars. Of course, people from Earth could not survive on Mars without some kind of adaptation. But the human body does seem to adapt, naturally, by increasing the amount of blood cells when necessary, over time. The people on Mars are a testament to that.

BUS

A study by Scott C. Warring of "UFOsightingsdaily.com" was a structure on Google Mars. The object looks something like a bus. It has a tapered, wedge shape, with front end smaller than back end. It measures roughly a mile long. There are windows along the sides. There are intermittent spot like tracks behind it, which suggest that the tires have protrusions on them for driving in deep sand. The enormous size must be to accommodate the gigantic Martians.

Bus on Mars
8/14/13 mgh

PART 5: EARTH-MARS EXCHANGES

NEMESIS (SUN-PARTNER)

The theory of the sun-partner Nemesis with its 26-million-year orbit has withstood the test of time for almost three decades now, (1984-2013). Yet, it is still controversial and debated, even today. Other reasons for the periodic extinction rate increases exist, while some scientists doubt the existence of the theoretical star, and even NASA considers it an unnecessary theory. Until and if Nemesis is found, it may never amount to more than one theoretical explanation of many explanations for the extinction of the dinosaurs. We need to be aware that the dinosaur extinction is just one of many familiar extinctions. If sun-partners are not to blame for periodic familial extinctions, then why would they be so periodic?

But from a completely different viewpoint, what can explain the existence of Earthlike animals and humans being found on Mars, from time periods 37.5 million years apart, other than from a collision being caused by a sun-partner? It must have been something big enough to change planetary orbits and therefore has to be larger than an asteroid or comet. The following time line was presented in "Life on Mars 2":

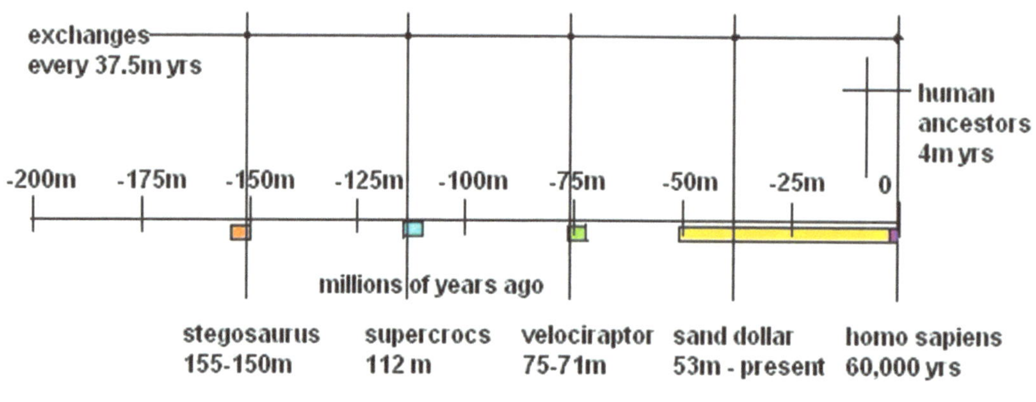

Time Line

Something managed to affect Mars' orbit more than Earth's, causing their collision. The theory about Nemesis is explained in Richard Muller's book, "Nemesis, The Death Star", (1984). Nemesis , with an orbit of 26 million years, and most recently estimated at its greatest distance from the Sun, 1.5 light years away. The closest known star to Earth is Alpha Centauri, 4 light years away. However, by the theory, its latest approach to our Sun was 11 million years ago.

So this other object, to summarize, whatever it may be, a planet or comet or asteroid or red or brown dwarf star, companion to our Sun, i.e., sun-partner, (some great mass orbiting our Sun), whatever it may be called, passed close enough to modify Mars' orbit, and caused Mars to collide with Earth, and, amazingly, did it on a regular basis, every 37.5 million years, and most recently, roughly 5 thousand years ago, in the time of Noah's ark. The following sketch shows how the orbits may have become elongated to cause collisions. Everything in the sketch is exaggerated and not to scale. The planets, as well as the Sun, due to gravitational attraction, would shift towards the Sun-partner, and their orbits would ellipticalize. Since the planets travel more slowly as they are more distant from the

Sun, the orbits of Mars, asteroids, and Jupiter would be affected to a greater extent than the orbit of Earth. Mercury, Venus and Earth would ellipticalize less, due to their faster speeds. Hence, Mars orbit would intersect Earths, while Jupiter's might have intersected Venus, as well as Earth, Mars, and the asteroid belt.

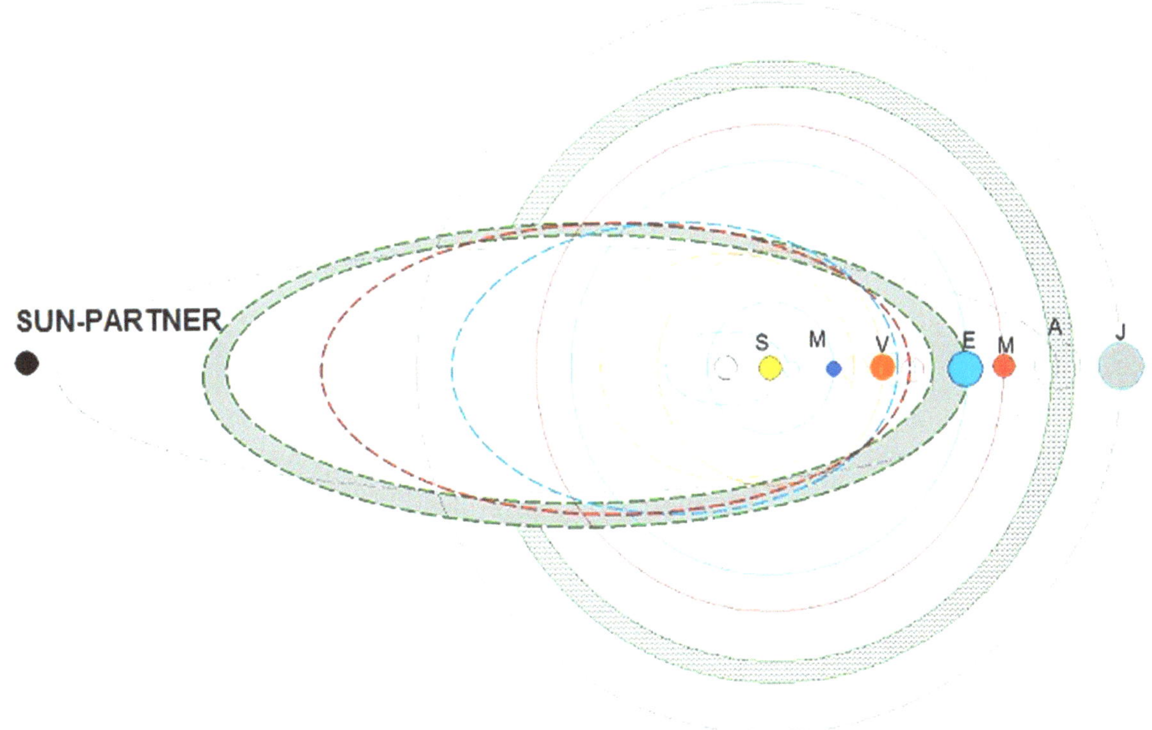

EXCHANGE THEORIES

If Mars' orbit was elongated, due to gravity from a second star nearby, then Earth's orbit would have been perturbed, as well. Therefore the points of exchange might have been at the sides of the ellipses. The two ellipses might intersect because Earth and Mars are of different mass and therefore would have different inertias and also speeds. Also the location of Nemesis with regard to the solar system is unknown. According to the Nemesis hypothesis, Nemesis passed outside the Oort Cloud far from Earth, but still, close enough to disrupt comet orbits, sending some close to Earth. According to the theory within this book, the asteroid belt became distorted as well, sending it into the path of Earth's orbit.

DIFFERENCES IN VELOCITY

When a sun-partner passes near the Sun, there would be, theoretically, considering simple physics, orbit shifts of all the objects in orbit about the Sun, as well as a shift in the relative location of the Sun itself, due to the gravitational attraction of the sun-partner and those bodies. The orbits would become elongated towards the sun-partner. That, alone, would not explain how the Earth and Mars would come to collide, since they might all shift equally. For planets to have collided, their orbits would have had to have shifted differently. Factors influencing orbit shifts are mass, inertia, distances from the Sun, and velocities. Galileo, dropping things from the top of the tower of Pisa, showed how objects of different mass accelerate the same due to gravity. Therefore, the difference in mass of asteroids, Mars and Earth may not have been the reason for their collision. A grain of sand orbiting the Sun would be drawn off its path the same as a massive object like a planet.

A bullet with a higher muzzle velocity will travel farther, and that fact implies that orbits shift, due to external forces, in relation to their velocities.

Therefore, the collision of Earth and Mars, as well as with the asteroids, might have occurred due to their differing velocities, which influenced their orbit shifts when the sun-partner passed nearby. The amount of shift in orbit is inversely proportional to the velocity of the orbiting objects, so that Earth's orbit would shift less than Mars' or asteroids'. Earth travels around the Sun at a speed of 66,630 MPH, Mars 54,000 MPH, and Jupiter 29,240 MPH. So the speed of asteroids in the asteroid belt is probably various, around 35,000-45,000 MPH. Therefore, the orbit shift of Earth would be less than the asteroids' and Mars' orbit shifts, making collisions possible during the time that the orbits intersect. During the Sun-partner pass-by of the Sun, it would cause increase in asteroid impacts on Earth and Mars, as well as possibly Venus and Jupiter. In fact, Venus and Jupiter might collide, which Velikovsky suggested in "Worlds in Collision".

WHEN DID THE FIRST HUMAN GO TO MARS?

The answer might be found by studying Martian gigantism/miniaturization. Assuming that the first humans to arrive on Mars, and did not realize that the conditions led to gigantism, they therefore likely lived many generations as they did on Earth, in houses on the surface. Let's say the super-giant humans are now 300 miles tall, average, based on the man with 90 miles shoulder span, and the little girl with 30 mile wide head, which suggests she is also 300 miles tall. Then, assume that time went by and another exchange occurred, and that exchange led to the people 60 feet tall. Then the third and last exchange brought the third group of normal sized people. The number of sizes of Martians today indicates the number of collisions between Earth and Mars in the time of homo sapiens. So then let's assume that the size of the offspring increases at a constant rate, with each generation, and let's assume that each generation takes 20 years before reproducing. Let's also assume that the normal sized humans learned to protect themselves by going subterranean. Therefore, they did not become gigantic. As for the tiny humans, the opposite occurred, due to some kind of magnetic-polar influence. The sizes of the people on Mars might be a clue to the time that

they have spent on the planet, as well as the amount of protection they had from cosmic radiation during their lifetimes.

THE GREAT FLOOD

Someone knew about the coming collision of Mars, before it happened, around 2800 BC, in advance. Noah was warned by someone. Astronomers of the time might have been capable of following Mars' path in the sky, and probably could have determined that it was on a collision course, if it were growing in size while remaining stationary in the sky. Someone aware of that could determine what was happening, within a few months of its colliding. But 120 years prior to the collision, it seems difficult to determine that a collision was to occur, and the thing an astronomer might notice is a previous collision or the appearance of a second Sun along with an understanding of the order of the solar system, orbits, and general concepts not available at the time. By tracking its path over many years or decades, the pattern would be known and projected for hundreds of years in the future. Regardless of who it was that warned Noah, he built his ark and it was filled with animals over a120 year period. Was it 120 years prior that the Nemesis first affected Mars' orbit? Perhaps the Martians saw it coming and Mars collided with Earth at that time, 2700 BC, approximately, and that was the first coming of the Martians to Earth. That is the time that the Watchers fell to Earth, (the male parents of the Nephilim). Noah was considered to be the best example of moral humanity, and considering that he lived for 800 years, he was quite a guy. Also, at the time, supposing that Martians and Earthlings were on both planets for some time, and communicated by some kind of space travel vehicles, the interchange seems more likely.

But then the question becomes, how and when did humans get to Mars? If there was a collision 37.5 million years ago, human development was non-existent, at least on Earth there is no record of such. So, how can we know the approximate time that humans first went to Mars? Perhaps it

will be discovered from fossil records on Mars, someday, or possibly, from communications with Martians, or perhaps, even, from fossils on Earth?

There seems to be a vague suggestion of evidence that Martians somehow came to Earth in the distant past, prior to the flood in the time of Noah. The Biblical accounts suggest that giant angels or godlike humans fell from the sky in the ancient past, prior to the great flood of Noah, and cohabited with humans, whose offspring, the Nephilim, ultimately died off or were killed. They lived in a society near what is now the Black Sea. In the times of Noah, a second group of Watchers came to punish them for violating rules that prohibit reproduction with humans. In fact, the origin of the Caucasian race is believed through DNA research to have stemmed from migrators from Africa, through India, to Europe around 60,000 years ago. Bones of giant humans have been found on Earth, petrified and embedded in coal deposits. Goliath, who was slain by a stone slung by David, was described as 9 feet tall, and was one of the last remaining Nephilim.

Did God actually go to Noah and tell him about the sun-partner and its affects? Or did he just say he's going to flood the Earth and how he does it is not Noah's business? Or, did God just give Noah the ability to realize what was going to happen, by his intelligence? Maybe Noah saw the approach of the Nemesis in his early youth, and then saw the two Suns, in middle age, and then saw the departure of the Nemesis and concluded that what happened in his youth would repeat. In his youth, possibly, a collision with Mars resulted in transfer of humans to Mars, who then grew to gigantic size, and then returned to Earth on the departure collisions. So perhaps there were two collisions or near misses in his childhood, 120 years apart, and then two in his elder years, and he experienced them all, so he knew that a flood was coming in 120 years after the first.

A THIRD SUN?

The incongruity between the 26 million-year cycle of extinctions and the 37.5 million-year cycle of Mars collisions suggests that there could be two Sun-partners: one that causes extinctions and another which simply causes orbit shifts and transfer of life forms between Mars and Earth, without causing extinction rate increases as large as during Nemesis passes. Or, possibly, if the two stars orbit the Sun in 26- and 37.5 million-year cycles, then the two would come together, both at the same time, occasionally, which occurred apparently around 37.5 million years ago, causing the great extinction at that time; which, incidentally, shows as the greatest extinction rate on the chart. The chart below was prepared by Richard Muller, in the 1980's, as basis for the Nemesis Theory at that time. Studying the chart below, there appears to be two significant periodic extinction rate increases, one every 26 million years, and another one every 37.5 million years. This suggests two sun-partners.

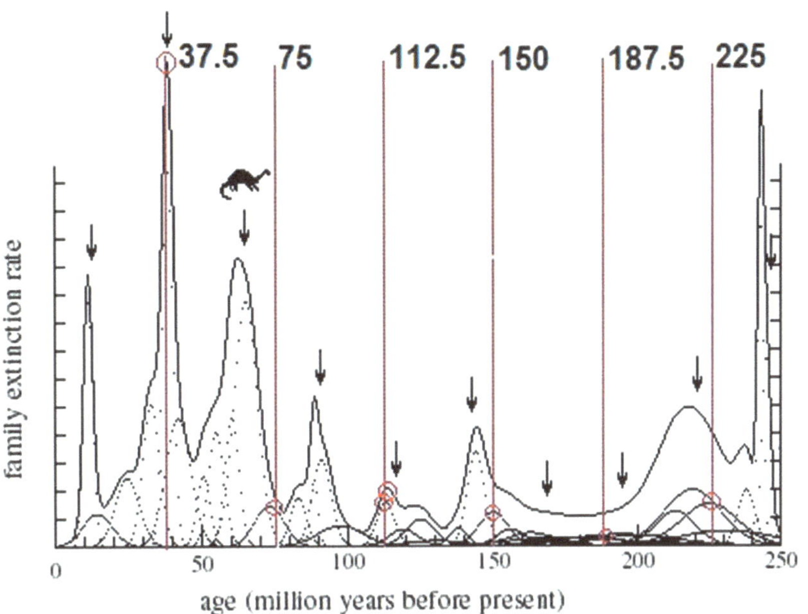

The Nemesis theory claims that the dinosaurs became extinct, (where indicated by the dinosaur image on the above chart), at the time of the passage of Nemesis close to the Sun, 65 million years ago. The extinction rate began to rise 10-15 million years prior to that, approximately 75-80 million years ago. The arrows on the chart indicate the 26 year period of repetition, which led to the Nemesis theory. It appears that a possible-second sun-partner initiated the extinction rate increase, during the extinctions, at 37.5, 75, 112.5, 150, and 225 million-year periodic occurrences, indicated on the chart with the red vertical lines. The two sun-partners seemingly worked together in elevating the extinction rate, like a one-two punch, one following the other over several millions of years.

A FOURTH SUN?

Studying the chart further, is there a third or even more periods of extinction, in addition to the two mentioned above? The question can be answered by looking at the remaining extinction-increase curves. In other words, if we ignore the two cycles above, what remains, and is it periodic? Consider the following chart:

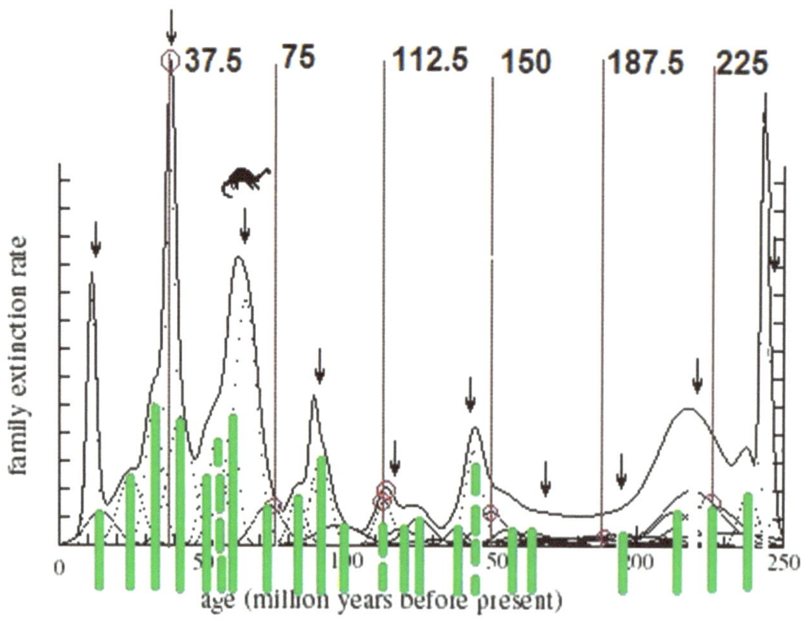

In looking at the extinction curves that occurred, ignoring those on the 26 and 37.5 million year cycles, (considering the green lines at the bottom of the chart), there appears to be a somewhat

erratic but also somewhat periodic extinction curve peaking at roughly 10 million year intervals, the last having occurred about 15 million years ago. Therefore it would seem that one sun-partner pass may have occurred unnoticably during the time of mankind on Earth, i.e. 5 million years ago. Perhaps there is a 10 million year cycle but sometimes it does not lead to extinction. Notice also a lull in the extinction rate increases during the 160-190 million-years-ago period, which could mean either that the extinction data is incomplete, or that there are times when the sun-partners do not cause extinctions. Rather, they pass harmlessly, possibly perturbing orbits but not enough to cause asteroid belt shift into Earth's path. And in fact, extinctions are occurring at the present time, such as the polar bear, tiger, and other mammals and animals, due to what most scientists call the global warming, caused by mankind, in theory. And of course there are various causes of extinctions, such as volcanoes, earthquakes, atomic bomb testing, and other similar environmental and manmade disasters. Mankind on Earth has been very fortunate in terms of sun-partner extinction rate increases.

WHERE WAS ATLANTIS?

According to Plato, (and we have little else to go by), Atlantis was an island civilization outside the Pillars of Hercules, now termed the Straits of Gibraltar. If the collision between Mars and Earth occurred at the Mediterranean Sea, (which was theorized in "Life on Mars 2", then an island outside of the Straits of Gibraltar would have been flooded many miles high with ocean water. It also might have been crushed by the planet Mars, as it careened across the Mediterranean and into the Atlantic. The path of travel would have taken the great flood across the Atlantic, and then to North America, roughly where the Gulf of Mexico now sits. And, continuing on, it would have brought the several-mile high flood across the land west of the Gulf of Mexico, and to the Pacific Ocean. Researchers have discovered a possible ancient sunken road, the "Bimini Road" along that path, near the Bahamas. So the location of Atlantis as described by Plato might have been in the direct path of the Mars-Earth collision.

THE COLLISION OF EARTH AND MARS

To help imagine what may have happened, when the Earth and Mars collided, we need to try to visualize two spinning tops, side by side, that bump into each other for a short instant. The Earth spins at 1000 miles per hour at the equator, while Mars spins at about half that speed. The tops, when they touch each other, tilt on their axis of rotation, momentarily, then the axes wobble for a short time, then eventually the axes return to their verticality.

Planets farther from the Sun travel more slowly in their orbits. That fact suggests that prior to the collision, Mars must have had an orbit closer to the Sun than it is today. It's velocity in its orbit around the Sun was therefore probably greater than it was today. Upon colliding with Earth, Mars' velocity was likely reduced, and its orbit distance from the Sun was increased, as it was "bumped" farther from the Sun, by Earth.

Earth's orbit, similarly, was likely farther from the Sun than it is today, and upon colliding with Mars, it probably was reduced in orbit diameter and increased in velocity, slightly, about the Sun.

Another reasonably expected result of the collision would have the decreased rotational velocity on axes, for both Mars and Earth.

The rotation of Mars is now similar to that of Earth's, which is roughly once per day. However, since Mars is smaller, the velocity is smaller. Prior to the collision, however, Mars velocity would have been greater than today's. The velocity of Mars at the equator today is roughly 500 miles per hour, whereas prior to colliding with Earth, it may have been double that. That is because the velocity change of Mars would have been much more significantly reduced by the collision than Earth's.

Earth's velocity would have been reduced by the collision. What we see today as 365.25 days per year, (the number of rotations in a year), may have been 360 days prior to the collision. In fact, ancient calendars were based on 360 days, for reasons unknown today. It could be that in ancient times, the 360 days was the number of days in a year and led to what we use today as 360 degrees in a circle. Why are there 360 degrees in a circle? Why not 100, or 400? How coincidental, that it is so close to the number of days in Earth's year. And where and when did the 360 degree circle concept originate? Likewise, why are there twenty-four divisions of the 24,000 mile Earth's circumference into hours? Why are there sixty minutes per hour, and sixty seconds per minute? Where did those numbers come from? Nobody knows.

Because the planets are spherical shapes, the effect is similar, but slightly different from the tops analogy. The moment of first contact between the two spheres causes the axes to rotate, and therefore the scratch on their surfaces is not a straight line, but a curve. When the axes try to straighten vertically again, the scratch again curves. Due to the wobbling of the axes, the net result, theoretically, becomes an "S" shaped scratch. This can be seen in the S shape of the Mediterranean Sea, and also in the Caspian Sea. These are possibly two places where The Earth was scratched by Mars. Of course, other possible collisions might have occurred on Earth's oceans, and those would not leave scratches. Nevertheless, there are two possible collision scratches on Earth, and one on Mars, that are easily visible.

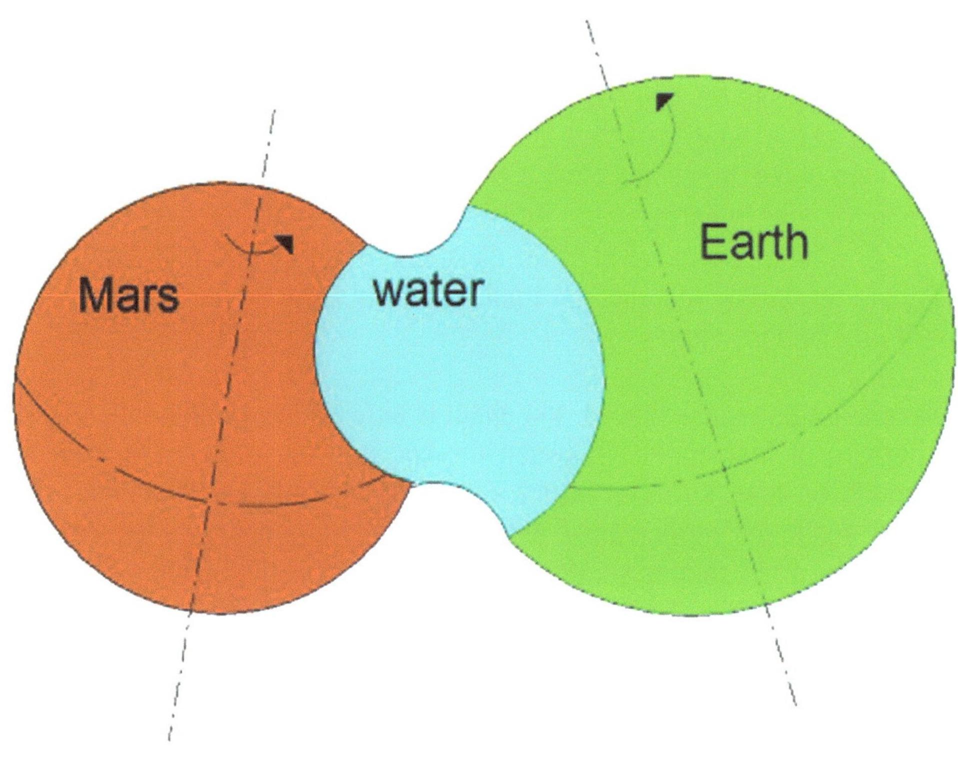
Mars
water
Earth

PART 6: EARLY MARS EXPLORATION

TESLA'S ELECTROMAGNETIC WAVE COMMUNICATIONS

Nikola Tesla, in 1899, claimed to have received electrical signals in a facility in Colorado, from unknown sources, which he believed were from Mars. However, he never published any of the details, and his claims are essentially unproven and ignored. His "teslascope" was a device that he used to make electromagnetic waves audible, so that he could listen to them. Tesla died in 1943. There is no record of his communications with Martians.

THE WOW SIGNAL

SETI (Search for Extra Terrestrial Intelligence) has no record of potential messages from Mars. However, there was one signal, considered as possibly extraterrestrial in origin, which is now called the "WOW! signal", detected on August 7, 1977, 11:16 pm from their observatory at Delaware, Ohio. The wavelength of the signal was precisely the wavelength that the researchers expected. The signal was detected from the region in the constellation Sagittarius. In analyzing this signal, one needs to consider that illusions occur in astronomy due to the speed of light. What we see is not the same as what truly is, because of the speed of light. Curiously, light takes 2.5 hours to get from Uranus to Earth, which is approximately the time it takes for rotation of the Earth on its axis 37 degrees. The signal was received when the receiver telescope was pointed near the location of the star Tau Sagittarii, at approximately the center of Sagittarius, about 37 degrees from Uranus' location at the time. The signal, therefore, might have originated as an Earth signal that bounced off of Uranus. The composition of Uranus' core is not precisely known and whether it could reflect electromagnetic waves is unknown.

PSYCHIC COMMUNICATIONS

Some people on Earth have claimed psychic communications with Martians. This was particularly popular in the 1920's, about the same time that Houdini's wife was trying to communicate with her late husband using psychics and Ouija boards. The communications that were claimed to be received indicated that Martians are people like Earthlings and have much interest in communicating with Earthlings. The concepts received generally indicate very positive, non-hostile Martian who are very much the same as Earthlings, physically, and are anxious to communicate with us.

MARS FLASHES

A Japanese astronomer, Tsuneo Saheki, observed Mars carefully during the period 1925-1962 and recorded his observations with drawings which are recorded on the internet. One observation that he noted was a brightening, or flare, in the Edom region, a pulse that lasted about five seconds, on July 1, 1954.

Brightness fluctuations were also detected from Mars' Edom Promontorium, in Shaparelli Crater for several days during June 7-10, 2001, from a team of United States amateur astronomers, in Nebraska and Florida. The flashes could have been natural reflections from ice or some highly reflective crystalline surfaces. The observations were defined as pulses lasting approximately 15 seconds, with three seconds of maximum brightness. Eight and ten-inch reflector telescopes were successful in viewing the fluctuations. The distance between Earth and Mars was about 35 million miles at the time. These observations are recorded on the internet.

GRAVITY WAVE COMMUNICATIONS

In 1988-89, Greg Hodowanec, an electrical engineer, doing research in gravity waves, believed he received intelligent signals, similar to Tesla's experience with electromagnetic waves, something like Morse Code impulses, from what he eventually also believed were from Mars. He ultimately, over several months, claimed to have had crude conversations with an entity which he concluded was on Mars. According to Hodowanec, the entity came to understand a few words, including "face", "man", "Earth" (as his planet), and "Mars" as the entity's planet. Hodowanec claimed the Martian entity had a word for the tenth planet. In 2005 a Trans-Neptunian dwarf planet was discovered, slightly larger than Pluto, and three times as far from the Sun as Pluto, now named Eris, also known as tenth planet Xena for a short time. The discovery of the object caused scientists to demote Pluto from "planet" to "dwarf planet". The Martian communicator confirmed to Hodowanec that Martians look like Earthlings, and also, Hodowanec derived from his conversations that Martians have colonized Earth in the past. Hodowanec received messages which seemed to duplicate his own transmissions, with letters A, C, E, K, N, R, S, T, which would have been Morse Code signals using gravity waves. It is surprising that the Martian with whom he was supposedly communicating actually was able to comprehend Morse Code, alphabet, and English language. Any communication with Martians in the future will require that either they learn an Earthly alphabet, language, and means to transmit it, or that Earthlings learn the Martian alphabet and language, or, possibly, both learn a new universal language. An educated guess about the language on Mars would be that it is similar to, derived from, or possibly branched off of Ancient Egyptian hieroglyphs and the language spoken in Ancient Egypt. Our alphabet is derived from the Phoenician alphabet which dates back to approximately 1200 BC, and therefore it did not exist at the time of the hypothetical Earth/Mars collision, in about 2700 BC, and hence, Earthlings who were transferred to Mars at that time would not have been aware of it. Martians might have developed methods of receiving Earth transmissions of television and radio signals. Hodowanec's research is recorded on the internet.

POSSIBLE FUTURE LIGHT SIGNALING

In early April, 2013, Mars passed behind the Sun in relation to Earth, which is termed "conjunct". This condition disrupts any communication NASA has with the rovers and satellites on Mars, and also it precludes any kind of electromagnetic or light wave contact with Martians. The conjunction effects last for a few weeks minimum, and occur regularly every two years plus a month. Coincidentally, Venus is passing behind the Sun at almost the same time during the April 2013 conjunction. The time to best view Mars flashes is not at the conjunction, but rather, prior to and after the opposition, while the Earth and Mars are closest together. To view the dark side of Mars from the dark side of Earth, the periods of roughly 2 or 3 months prior to and after the opposition will be best for light signaling. A possible method to pursue would be using lasers from the dark side of Earth to the dark side of Mars, which would allow several hours of transmission and reception daily. It would be necessary to observe Mars in a telescope while signaling. Whether Martians have lasers is unknown. Another possible method would be to use mirrors to reflect the Sunlight. An attempt of this type of signaling should consider light speed, with Mars' true position being a few minutes off of its apparent location, and so, beaming a light to Mars would be similar to hitting a moving target.

The planet Mars' elliptical orbit brings it 128 million miles from the Sun at its closest point, and 154 million miles at its farthest. The planet Earth's orbit brings it 94 million miles from the Sun at its farthest point and 91 at its closest. Therefore, proximity of Earth and Mars point varies over time, from roughly 34 million (128 minus 94) miles at the closest point to over 230 million miles at the farthest. During early April of 2013 Mars and Earth were farthest apart, one AU (distance Earth to Sun) plus 1.5 AU (distance Sun to Mars), or totally, 2.5 AU, over 200 million miles. The closest approach is the time when light communication between Earth and Mars (dark sides) would be most likely successful. The next time period when that will occur is roughly early to mid-2014, and then every two years plus a month afterwards. June of 2001, the date of the observation of Mars

"flashes", was a similar time, when the two planets were most closely together. In fact in 2003 the two planets were closer than they have been in 50,000 years; (not counting the possible collision during the Great Flood around 2700 BC. and during previous Earth/Mars collisions, as discussed in Part 5.)

RUSSIAN PROBE

Russian attempts to send rovers to Mars, beginning in the 1960's, all failed, some at the last moment. On March 25, 1989, an elliptical-shaped projectile shadow was seen approaching the Russian spacecraft "Phobos !!" as it entered the Martian atmosphere. In the photograph below, the shadow of the object was seen. It may have been a man-made projectile or a natural object such as a meteor. Whatever it was, it apparently collided with the Russian probe and disabled it, just after the photograph, causing loss of Russian contact and failure of the mission. Did the Russian craft get shot down, or was it hit coincidentally by some small orbiting satellite? The answer is still not known.

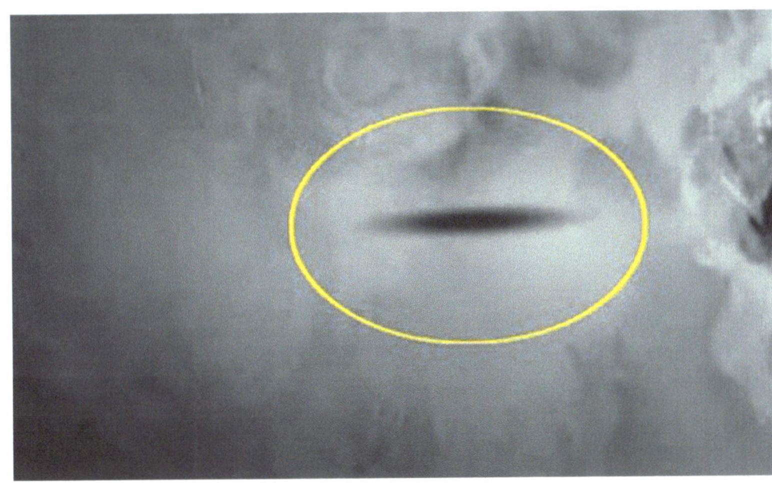

CONCLUSIONS

CONCLUSIONS IN PRIOR BOOKS

The first book, “Life on Mars”, determined that indeed there is life on Mars, and that there were traces of humans on Mars; photos of humans were questionable, not clear enough to conclude outright that humans existed on the planet. Conclusions were as follows:

- There is a need to explore the various scientific disciplines regarding the life forms on Mars, such as zoology, biology, paleontology, botany, and other life sciences.
- A simple test chamber could be made and animals put inside with regulation of the atmosphere, to see how close they can come to surviving in an atmosphere similar to Mars’

The second book, “Life on Mars 2”, clarified how there were indeed humans on Mars, of various cultures and sizes. These revelations led to the following conclusions.

- We need to study the people and societies on Mars, their cultures and their history, as well as their anatomy, their anthropology, sociology, psychology, and whatever other scientific areas we can.
- We need to learn more about the Martian language and we need to learn how to communicate with them.

CONCLUSIONS IN THIS BOOK

Major discoveries of this third book included the discoveries of the tiny people on Mars, as well as the theoretical determination and/or postulation of additional Sun-partners. These new discoveries lead to the following conclusions.

- Communication with Martians could completely rewrite our objectives regarding the planet. If humans exist on Mars, why are we trying to send researchers there at such great expense? Whatever they need to do, the Martians could do as well with communication, teaching, and learning.

- The current plan of NASA is to study geology and ancient water flow as if they are totally ignorant of animal and human life on the planet. Knowing that humans live on Mars, wouldn't it be better to bring Martians to Earth? If we are planning to bring back a dirt sample from Mars to Earth, why not bring back a Martian? A tiny Martian, one inch tall, would be easier to take to Earth than it would be to take a huge 150 pound human to Mars.

- This cursory review shows that we may have many Sun-partners, possibly four or more. Some cause extinction rates to rise, while others sometimes do, but not always. Sun-partners need to be confirmed and examined by visual observation or other means. Hopefully they will be found soon.

ON SENDING PEOPLE TO MARS

A Netherland group plans to send people to Mars in 2023 (Mars 1). NASA plans to send people in the 2030's. At this time, a trip to Mars takes roughly nine months. It can be accomplished with our current technology, but at great expense. Furthermore, the Netherlands group, four individuals initially, would not return to Earth due to the extreme cost. Is this really what we should do?

- A typical human on Earth requires roughly 2000 calories per day of food and about a quart of water. Over a nine month period, that amounts to roughly 1000 pounds of food and water, although it seems likely that it might be cut in half with some careful planning; by limiting activity, induced sleep, weight loss, or whatever. In addition there is an equal amount of waste material to deal with. By comparison, a tiny Martian would require a small fraction of that. If we take the proportionate volume, (a typical human would have a volume of roughly two cubic feet, while a tiny Martian would have more like two cubic inches). The ratio is 4300 cubic inches to 2 cubic inches, or 2650 to 1. If the weight of food for a normal human is 1000 pounds, then the weight of food for a tiny Martian would be 1000/2650 or .37 pounds, (roughly half a pound). The rocket fuel needed for transporting the tiny person, including the rocket fuel to transport the food and water and the rocket fuel for the rocket fuel, would be very much reduced, and far more practical, cost saving, and achievable.

- The cost of travel to Mars is huge. Nevertheless, with our current system of monetization, we can essentially afford almost anything we want to. After all, the wealth on Earth is simply psychological; like paper money, coins made of abundant metals, and bonds, it is based on promises and optimism. As long as the optimism remains, the wealth will remain. However, out of practicality, if there is a lower cost alternative, we should take it.

- The fact that humans already exist on Mars is evidence that acclimation is feasible. We need to test this concept thoroughly in order to assure ourselves that we can do it. Tests should be made on mice and other animals first, and finally, with humans, to determine if they can adapt to the environment on Mars. In reality, that is probably what happened previously. But also, with humans on Mars, perhaps we do not need to send Earthlings to Mars. Anything we want can be attained by those who are already there, provided that we establish communications.

GOALS

These books are a method by which I am able to record my observations and thoughts, and to communicate them to others. But beyond that, they are a means to establish a new order of understanding amongst fellow Earthlings about our lost past and our inability to retrieve it. Our goal in the future must be to pursue truth and knowledge, as we have always done. However, we also must accept that we are limited and our perspective is narrow. We must question our beliefs and replace them with truths and facts where we find them necessary. We must be open to new realities, everywhere.

We must face the reality that we have long-lost relatives on Mars, and we must come to an understanding of that. We must take new steps to regain lost knowledge such as the myths of giant gods and pixies. And finally, we must apply our newfound knowledge to exploration and discoveries. to yield ever more fascinating truths.

www.ingramcontent.com/pod-product-compliance
Ingram Content Group UK Ltd.
Pitfield, Milton Keynes, MK11 3LW, UK
UKHW060111300726
14090UKWH00002B/130

* 9 7 8 1 4 8 3 6 8 4 3 8 3 *